Alexander B. Johnson

Deep Sea Soundings and Explorations of the Botton

Or, the ultimate analysis of human knowledge

Alexander B. Johnson

Deep Sea Soundings and Explorations of the Botton
Or, the ultimate analysis of human knowledge

ISBN/EAN: 9783337248536

Printed in Europe, USA, Canada, Australia, Japan

Cover: Foto ©berggeist007 / pixelio.de

More available books at **www.hansebooks.com**

DEEP SEA SOUNDINGS

AND

EXPLORATIONS OF THE BOTTOM;

OR, THE

𝔘ltimate 𝔄nalysis of 𝔥uman 𝔎nowledge.

——— —

BY A. B. JOHNSON.

———

" Whence and what are we? —
What means the drama by the world sustained?"

COWPER.

———

PRINTED FOR PRIVATE DISTRIBUTION.

1861.

BOSTON:

PRINTED BY JOHN WILSON AND SON,

22, SCHOOL STREET.

CONTENTS.

CONTENTS.

DIAGRAMS.

DEEP SEA SOUNDINGS.

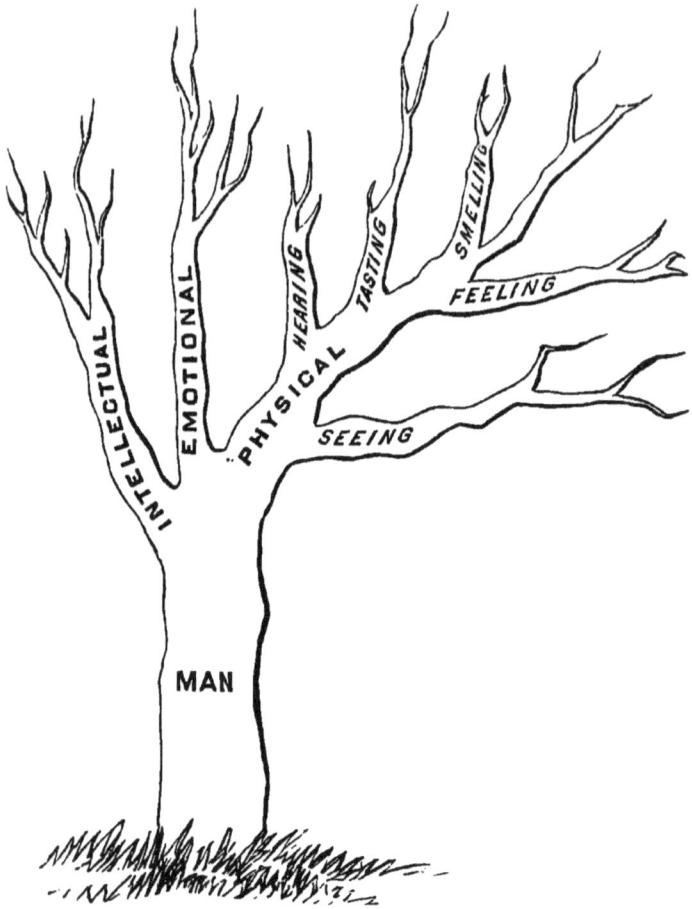

THE TREE OF KNOWLEDGE OF GOOD AND EVIL,
PREVIOUSLY TO ITS GERMINATION.

I.

MAN'S TRIPLICITY.

MAN is intellectual, physical, and emotional, as exhibited in the foregoing diagram : hence education is divided into physical, intellectual, and moral. Physical education relates to all that the senses of a man can perceive; intellectual, to all the processes of his intellect; and moral education, to all his emotional feelings. The most illiterate person is practically conscious of his triplicity as above; and he will say, "I *heard* a noise ; I *saw* a color ; I *tasted* an acid : " and he will discriminate equally well the two other senses when he is referring to their information. With like accuracy, he will attribute to his emotional organism the feelings which pertain thereto, and say, "I *feel* hopeful ; I *feel* angry, envious, proud, discontented, humble," &c. : and, when he speaks of his intellectual processes, he will employ words designative of his intellect; as, "I *think* I met A yesterday ; I *remember* his appearance ; I *imagined* he did not see me ; I *reflected* afterwards on the occurrence; I *judged* he was not well; I *conceive* he will be sorry; " &c. A nice distinction exists between the words " shall " and " will ; " *will* relating to the emotional organism, and *shall* to the intellectual. The misplacement of these words never occurs to an Englishman ; but a common joke attributes the misplacement to a Frenchman, who, when struggling in the water to prevent being drowned, exclaimed to the bystanders, " I *will* drown, and nobody *shall* help me." We often say, a man compliments his heart at the expense of his head. Everybody understands that the heart denotes the emotional organism ; and the head, the intellectual. " My heart is sad, but my head is ignorant of any cause therefor," refers to an emotional depression, and an intellectual ignorance of its cause. Should I say, " My head is sad," the assertion would not be understood as denoting an emotional depression, but a physical pain, or a diseased confusion of thought. These distinctions in the employment of words form no part of literary education ; but every man refers as spontaneously to his respective three organisms as he refers to his respective physical members, and says his *feet* shall carry him, his *hands* shall support him, his *back* shall bear the burden, his *tongue* shall speak, he will pour drink *down* his throat, and put meat *into* his stomach, &c.

Man's three organisms are recognized further by diseases peculiar to each; as, physical, intellectual, and moral ; moral being a synonyme of emotional. Idiocy, also, is a recognized condition of either a total or partial absence of the intellect alone, or of the intellect and emotions, while the physical organism may be perfect. Sleep affects the physical organism more than the intellectual, whose actions, under such a condition of the body, are called " dreams ; " and to these the emotional organism responds almost as vividly as during physical wakefulness. Old age affects the physical and emotional organisms earlier, usually, than the intellectual. Alcohol, taken into the stomach, affects the emotional organism first, then the physical, and last the intellectual. Whether anæsthetics impair alike the three organisms, I am not informed. Paralysis may affect all the organisms equally ; or one or two may be affected, and the others not ; and the organisms which are simultaneously affected may suffer in different degrees. The intellect was not involved in the curse denounced on man's disobedience. Emotional sorrow and physical subjection alone were imprecated on the woman ; and on the man, emotional sorrow and physical labor.

The triplicity of man impresses itself on our literature. Poetry and theology may be termed " the literature of our emotional organism ; " philosophy, astronomy, mathematics, and law, " the literature of our intellect ; " and treatises on trade, manufactures, arts, navigation, war, medicine, and surgery, " the literature of our physical organism." Men like poetry in proportion as their emotional organism preponderates over their intellectual; they delight in philosophy as their intellectual organism preponderates over their emotional; and they delight in manual operations as their physical organism preponderates over their intellectual and emotional parts. Our triplicity accounts for a triplication in phraseology that is too common to be casual; as, for instance, our forefathers pledged to the success of American independence " their lives, their fortunes, and their sacred honor." Honor was a pledge of their intellect; fortune was a pledge of their physical possessions; and life typified their emotional organism. In the Episcopal marriage ceremony, the bride promises " to love, honor, and obey " the husband. " Love " refers to her emotional organism; " honor " refers to her intellectual appreciation of him; and " obey " refers to physical services. The bridegroom pledges, in return, his three organisms; but, instead of physical obedience, he promises physical cherishment. Such triplications are numerous in the Scriptures; as, " What doth the Lord require of thee, but to do justly, and to love mercy, and to walk humbly with thy God ? " — " To do " refers to physical conduct; " to love " refers to emotional feelings; and " to walk humbly" refers to an intellectual abasement of ourselves. Again: " Let not once be named among you, neither filthiness, nor foolish talking, nor jesting." We find, by the context, that " filthiness " means impure emotions; " foolish talk " is physical; and

" jesting " is intellectual frivolity. Again : " Let every man be swift to hear, slow to speak, slow to wrath." The first refers to the intellect; the second is physical; and the third, emotional. The Episcopal liturgy deprecates " the world, the flesh, and the Devil." The " world," in scriptural metonymy, signifies physical possessions ; " the flesh," emotional enticements ; and " the Devil," intellectual suggestions. The expression, " I pray and beseech you," would be tautological : but " to pray " addresses the intellect ; and " beseech," the emotions. When King Hezekiah was informed that he should die, he said, " I beseech thee, O Lord ! remember how I have walked before thee in truth, and with a perfect heart, and have done that which is good in thy sight." Hezekiah's perfect heart is his emotional purity ; to walk before the Lord in truth is intellectual rectitude ; and to do what is good is physical rectitude. David said to Jonathan, " What have I done ? what is mine iniquity ? and what is my sin before thy father, that he seeketh my life ? " — " What have I done ? " is physical action ; " what is mine iniquity ? " is intellectual intention ; and " what is my sin ? " is emotional perversity. So Elkanah said, " Hannah, why weepest thou ? why eatest thou not ? and why is thy heart grieved ? " Grieved is emotional ; eating is physical action ; and weeping denotes intellectual reflections. Man's triplicity may also have relation to the thrice-repeated invocations that are numerous in the Bible ; as when the Prophet

Elijah stretched himself upon the widow's dead son three times, and each time said, "O Lord, my God! I pray thee let this child's soul come unto him again." Also when the Lord called Samuel three several times, while Samuel was sleeping in the temple; Samuel each time awaking, and supposing he was called by Eli. Again: when Azariah sent a captain with fifty soldiers to bring Elijah to him, and Elijah called down from heaven fire that destroyed the captain and his fifty. Then Azariah sent another captain with fifty soldiers, and they were destroyed in like manner. Then Azariah sent the third time a captain with fifty soldiers; and Elijah submitted, and went with them. So, when Elijah was to be taken up alive to heaven, he thrice attempted to escape from Elisha, &c.; and the oft-repeated "Alas! alas! alas!"— "Oh! oh! oh!" Possibly, as the globosity of our eyes shapes the field of our vision, our triformity originates triplicates; as, present, past, and future; heaven, earth, and hell; war, pestilence, and famine; life, death, and immortality; earth, sea, and air; beginning, middle, end, &c.: especially as each of our three organisms usually supplies a meaning to one of the three terms; as, in the foregoing, "present" time is physical performances; "past" is intellectual recollections; and "future" is emotional expectations; — "life" is physical action; "death" is emotional extinction; and "immortality" is an intellectual conception; — "heaven" is an intellectual conception; "earth" is physical; and "hell" is emotional, &c.

Man's triplicity impairs not his oneness; the two conditions belonging to different organisms. The oneness is a conception of the intellect which deems the emotions and the physical members as parts of a mysterious *ego* that intellectually thinks, imagines, conceives, judges, supposes, guesses, dreams, reckons, calculates, recollects; and emotionally hopes, fears, loves, envies, believes, doubts, grieves, covets, disbelieves, regrets, hates; and physically sees, hears, feels, tastes, smells, laughs, cries, jumps, walks, dances, hops, fights, sleeps, eats, speaks, &c. The Scriptures declare that man was created in the likeness of God; and this is assumed to mean man's physical form, though such an assumption is incompatible with God's filling all space, and with the adaptation of man's physical form to only definite purposes. The likeness finds a better solution in man's triplicity and God's tri-unity. "No man hath seen God at any time," and all his attributes are intellections; as, immortal, omnipotent, omniscient, omnipresent, uncaused, uncreated, without beginning and without end; a Being to whom "a thousand years are as one day, and one day as a thousand years;" and to whom "darkness and night are both alike." God the Father is, therefore, all intellect: but God the Son is physical, being manifested in the flesh, — a man born of woman; eating, drinking; being crucified, dead, and buried. God the Holy Ghost is all emotional. He is the Comforter; producing hope, joy, faith, gladness, and belief in the hearts of man. The three are one.

"And the Lord God took the man, and put him into the garden of Eden to dress and keep it." This employed his physical nature. His emotional nature (his forbearance and self-denial) was also exercised : for "the Lord God commanded the man, saying, Of every tree of the garden thou mayest freely eat; but of the tree of good and evil, thou shalt not eat of it." Nor was the intellect of Adam unemployed : for " the Lord God formed every beast of the field and every fowl of the air, and brought them to Adam to see what he would call them ; and whatsover Adam called every living creature, that was the name thereof." Man's intellect continues this employment ; and his emotions he names "love, hope, anger, joy, envy, jealousy, belief, faith, unbelief, vanity, pride," &c. His intellections he names " thought, memory, idea, recollection, judgment, knowledge, conception, contemplation, causation, effectuation, power, time, number," &c.; and to his sensible perceptions he has given names innumerable. The following diagram may denote more clearly the triple division of our knowledge : —

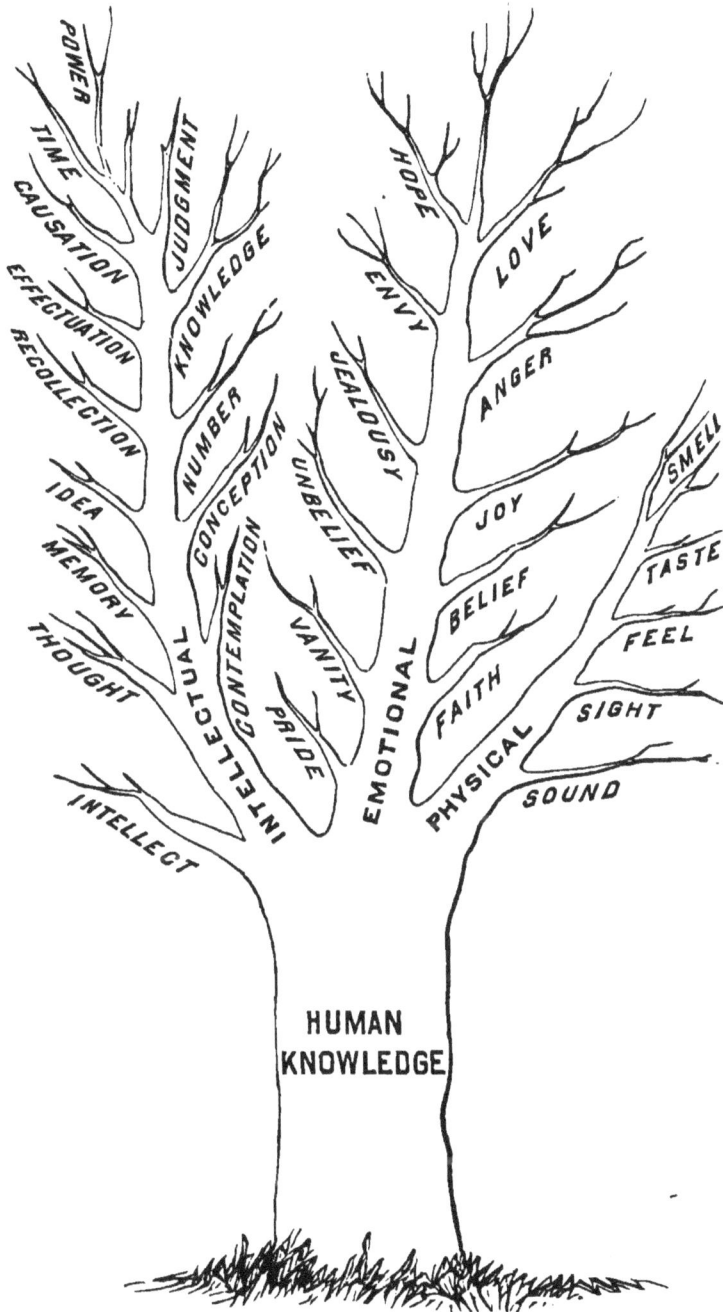

THE TREE OF KNOWLEDGE OF GOOD AND EVIL,
WITH SOME OF ITS GERMS AND FRUIT.

THE TRIPLICITY OF HUMAN KNOWLEDGE.

Possessing three organisms, and deriving through them all he knows, man's knowledge is tri-form. This might have remained undiscovered, if Providence had not manifested, that specific organic defects, as blindness, deafness, and idiocy, occasion specific deficiencies of knowledge. All that a man sees, tastes, smells, hears, and feels, corporally, constitutes his physical knowledge ; all he feels emotionally constitutes his emotional knowledge ; and all he knows otherwise is intellectual. If he take the wings of the morning, and fly (in thought) to the uttermost bounds of the sea, he is still within the magic barriers of the intellect, the emotions, and the senses ; if he ascend (in thought) to heaven, or descend (in thought) to hell, the same barriers are around him. He may enlarge the quantity of his knowledge by employing his senses in new directions, his intellect in new speculations, and his emotions in new experiences ; but he can gain no new kind of knowledge, nor transmute what is intellectual into physical, or what is physical into intellectual, or what is emotional into any thing but emotions.

The heterogeneity of our knowledge as above, and our indiscrimination of the heterogeneity when we speak, have ever perplexed speculation, but not

our practices ; man's three organisms continuing
their heterogeneous ministrations uninfluenced by
our speculations. For instance, Hume's speculative
denial of power, causation, and effectuation, pre-
vented not his emotional organism from responding
fear to any object that possessed power to cause
danger. The speculative scepticism of Pyrrho oc-
casioned a foolish tradition, that, to save him from
physical destruction, he needed friends to forcibly
remove him from danger : but a far more vigilant
and authoritative monitor in all such cases was his
emotional organism ; and it was ever present, and
incapable of being silenced. The speculations of
Hume and Pyrrho, being a product of their intellect, satisfied it ; and each philosopher's emotional
organism responded faith and belief to the speculations as logical conclusions, notwithstanding it responded fear in all cases of practical danger. This
evinces that our nature is adapted to both logical
speculation and physical practice, even when they
conflict ; and that neither can eradicate the other :
but, till a man understands the threefold heterogeneity of knowledge, his intellectual speculations become often logical puzzles. For example, the oneness
of a man, and his inability to find the oneness physically, is a great puzzle till a man knows that the
oneness is intellectual, and not physical. Man's
identity through infancy to old age is another puzzle,
when it is sought physically ; but the mystery ceases

when we know that the identity is not physical, but a conception of the intellect. The spirit, also, that animated Thomas before his death, and which no scrutiny can at any time detect physically, constitutes a puzzle till we know that the spirit is a conception of the intellect, and not physical. Time is another puzzle : our actions, thoughts, feelings, and sensations are all successive ; and, in contemplation of this organic condition, the intellect conceives time. We therefore need not wonder at our inability to find time physically ; or that, with an omniscient and omnipresent God, "a thousand years are as one day, and one day as a thousand years ; " or that, at the termination of the world, and all the incidents from which our intellect conceives time, an angel is to stand with one foot on the earth, and another on the sea, and, lifting up his hand to heaven, is to swear "that there shall be time no longer."

Our emotional organism, also, is the occasion of mysteries. "Man is born to sorrow as sparks to fly upwards ; " and he is equally born to believe and disbelieve, to hope, fear, love, hate, envy, revenge, and to numerous other emotional feelings to which our intellect has given names, and which, when thus personified, seem like little sprites located in the head or heart, but mysteriously undiscoverable physically while we are alive, and by dissection after death ; and for the sufficient reason that they are not physical. Properly discriminated into its organic classes, all our knowledge is free from mystery, or equally mysterious if we prefer to esteem it mysterious ; but we add an unnecessary mystery at our inability to discover sensibly what is not sensible. Human knowledge can be analyzed no further than into the organisms to which it pertains ; and all attempts to delve below or beyond this boundary are founded in ignorance of the inconvertibility, into each other, of the information yielded by our several organisms.

Chemistry finds that nearly all material substances are compounds, and it analyzes them into elements. Knowledge, on the contrary, is composed of elements, — sights, sounds, tastes, feels, smells, emotions, and intellections ; and the intellect compounds them into ideas, words, phrases, sentences, &c. We analyze physical compounds when we would understand their elementary ingredients ; so we must analyze intellectual compounds when we would understand their sensible, intellectual, and emotional ingredients : and thus only can we relieve ourselves from the speculative puzzles like those we have been considering, and which arise from deeming sensible what is either emotional or intellectual. For further information on this analysis, I refer to my two books, — "The Meaning of Words analyzed into Words and Unverbal Things, and Unverbal Things classified into Intellections, Sensations, and Emotions," published in New York, by D. Appleton and Co., in

1854 ; and "The Physiology of the Senses, or how and what we See, Hear, Taste, Feel, and Smell," published in the same city, in 1856, by Derby and Jackson.

DIAGRAM B.

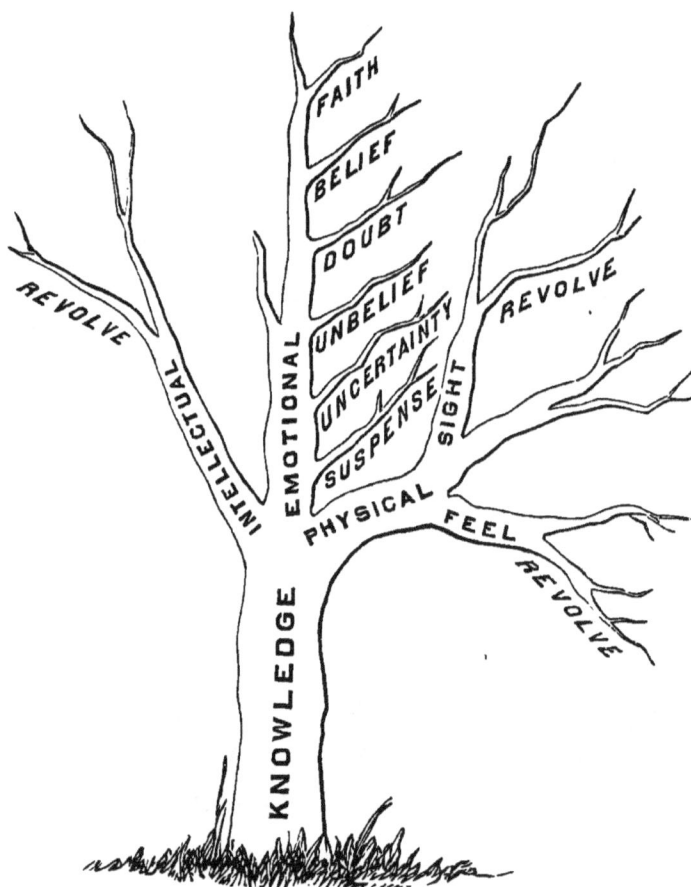

To exhibit, however, a specimen of the analysis intended, let us admit that the earth is constantly revolving from west to east. The word "revolve" analyzes (as displayed in Diagram B) into a sight, a feel, and an intellection. An artificial globe can be made to exhibit the sight revolve and the feel revolve: but the earth exhibits only the sight revolve; and yet your intellect conceives, that were you placed (as angels may be) externally of the earth, as you are placed externally of the artificial globe, you could feel the earth revolve as you feel the artificial globe revolve. To this intellectually conceived identity of the two revolutions, your emotional organism may respond with a feeling of belief therein exempt from any counter-feeling of unbelief or doubt; though some persons may have simply a feeling of faith that the two revolutions are physically identical, while other men may feel doubtful, and others may feel unbelief in the physical homogeneity of the two revolutions. Anyway, the analysis enables you to contemplate the earth's revolution in its organic diversity, relieved from intellectually conceived identities with other revolutions; and in no science is such an analysis so necessary to allay ignorant wonder as in astronomy.

Now, suppose, by way of proving the physical identity of the earth's revolution and the revolution of an artificial globe, you adduce any number of concurring sensibly perceived facts, and found thereon any number of intellectual conceptions; yet, as the feel revolve must ever remain unfelt in the earth, the first analysis remains unaffected, except as your new facts, &c., may influence my emotional organism

to believe what otherwise I might not believe. But
no belief can change physical facts; the information
of each of our three organisms being supreme within
its own intelligence, and each constituting a *quasi*
world which is a *terra incognita* to the other two
organisms: "around the things of each a great
gulf is fixed, so that the things of one organism, that
would pass to another organism, cannot." Finally,
we may say of human knowledge, as St. Paul says
of flesh, " All flesh is not the same flesh : but there
is one kind of flesh of men ; another flesh of beasts ;
another, of fishes ; another, of birds." So there is
one kind of knowledge of the senses, another knowledge of the emotions, and another of the intellect :
but they are all equally our knowledge ; just as
laughing and crying, sleeping and waking, dreaming
and acting, living and dying, are all equally our
humanity. We are, however, so constituted, that
our sensible perceptions are the most satisfactory
part of our knowledge, especially those that pertain
to the corporal sense of feeling. These are literally
real, substantial, tangible, solid, fixed, and so forth ;
but, when we know that these words name perceptions of the sense, of feeling, we need not wonder
that no other sense can yield us information that is
real, substantial, solid, fixed, &c., any more than we
need wonder that no sense but tasting can yield us
the knowledge of sweet and bitter. Still, our emotional organism responds certainty and confidence in

a greater degree to our corporal feelings than to any other parts of our knowledge. This is a psychological fact, which our intellect may speculate about, but cannot change.

THE TRIPLICITY OF THINGS KNOWABLE.

The Mammoth Cave of Kentucky, where light never penetrates, contains waters inhabited by fish that are formed without eyes. A like economy of creative power pervades the universe; aquatic animals being formed without feet, and terrestrial animals without fins : hence the possession, by man, of inlets for three essentially different kinds of knowledge, might alone lead a Cuvier. of some other planet to assume that our world includes three different kinds of phenomena, — a sensible world, which can be seen, heard, felt, smelled, or tasted, but known no other way; an emotional world, that can be felt emotionally, but known no other way; and an intellectual world, that can be recognized by the intellect alone. In contemplating even my own person, my senses can only perceive therein all that is sensible ; and an anatomist, who should cut my body into pieces, can only increase thereby the number of his sensible perceptions. None of the foregoing are recognized by my emotional organism; but it recognizes, in my person, hope, fear, love, hatred, desire, envy, rage, expectation, faith, belief, reverence, piety, scorn, pride, disdain, avarice, liberality, selfishness, patriotism, cheerfulness, despondency, lust, satiety, courage, and numerous other feelings to which names have been given, and many feelings which have received no name. My intellect recognizes, in my person, things different from any of the two foregoing. It discovers that all my sensible exhibitions are only *properties*, or *qualities*, of a substratum which the senses cannot see, but which constitutes my proper physical personality: and, though this personality is mortal and perishable, my intellect discovers it to be animated by an immortal soul, to which my intellect itself belongs, and my emotional feelings; and that the perishable body and imperishable soul make together one being, who has always been the same person from infancy to old age, in sickness and in health, in love and hate, in honor and dishonor, sleeping or awake. My intellect discovers also, that I, this mysterious unit, bear various relations to my Creator, to society, to my neighbor, parents, friends, wife, children, servants, country, &c.; and that to each of these relations I owe different duties, which require some of my emotional feelings and bodily actions to be restrained, some to be stimulated; and that, to accomplish these and other purposes, I possess a circumscribed power of causation and effectuation; and that I am only one of a numerous race of beings similar to myself, who with other beings, animate and inanimate, infinite in number and in variety of form and habits, occupy severally a limited portion of space, but only for a brief period of time, which is momentarily converting a portion of an illimitable future into present, and the present into an illimitable past.

The three worlds of which I have thus spoken are revealed to every man by only his own organisms. Through each of his five external senses enter all the items of the sensible world that come within the purview of the respective senses: hence men who inhabit the same locality possess mainly the same sensible knowledge. Man's emotional knowledge is experienced in common by all persons to a greater degree than his sensible knowledge; for that varies with localities, and may vary in different periods of time: but his emotional knowledge is unvaried by time or place; and men differ therein from each other only in degrees of its intensity, and in the frequency with which different individuals indulge given emotions. Our intellectual knowledge, like our emotional, is common to all men, in all times and places, so far as relates to its kind. All men, for instance, know power, causation, effectuation, quality, place, time, figure, relation, &c.; but the same items or manifestations of power, causation, &c., are not known to all men, and are infinite in

number and variety. Diagram C exhibits some of
the kinds of intellectual knowledge that are common
to all men.   They are thus common, by reason that,
everywhere and at all periods, circumstances exist
that excite their conception by the intellect ; just as
circumstances everywhere exist to excite the emotions
of hope, fear, love, hate, &c., and to excite the sensible perceptions of tastes, colors, odors, sounds, &c.
And as the eye needs no previous instruction to
perceive colors, or the emotions to feel fear ; so the
intellect needs no previous instruction to conceive,
under given circumstances, all the kinds of intellections that pertain to its organism.

DIAGRAM C.

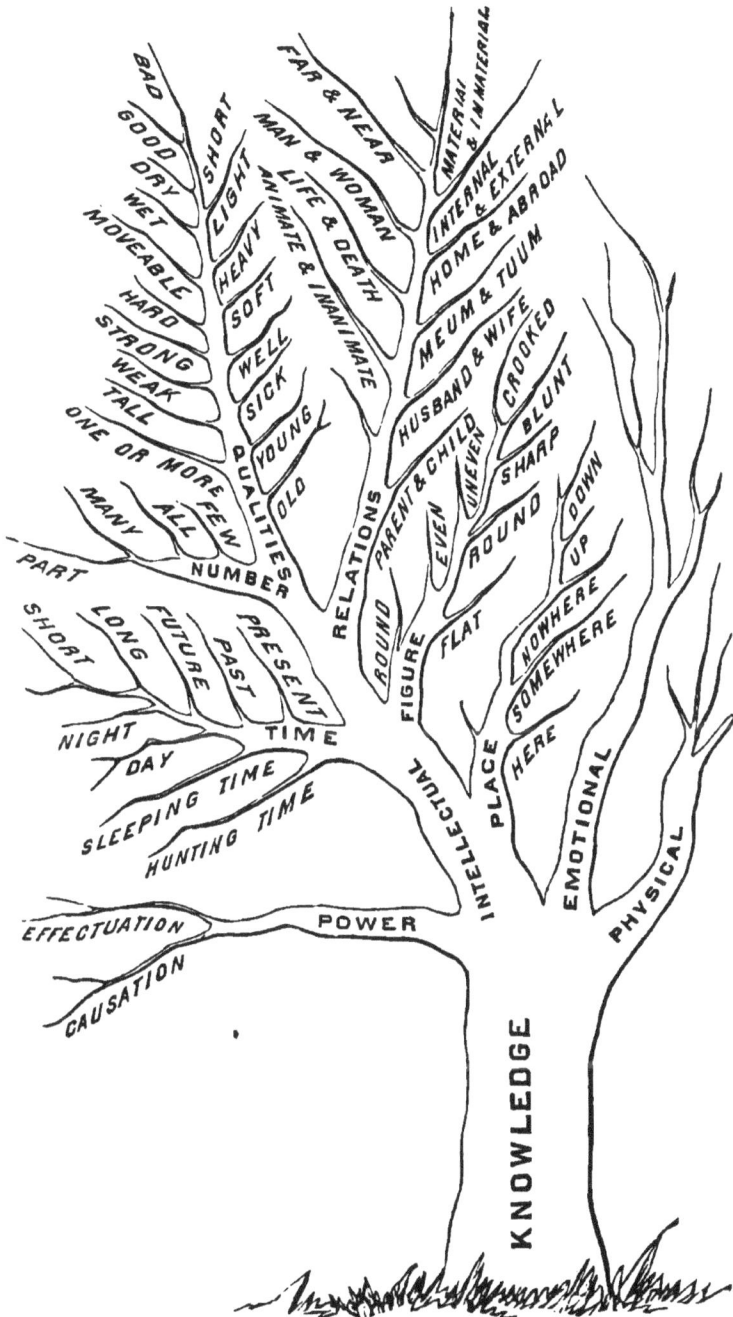

II. ·

THE ANALYSIS AND BOUNDARIES OF IDEAS ; OR, THE CYCLES OF THE IN-TELLECT.

As man can articulate only about forty different elementary sounds, the alphabets of all languages are
necessarily much alike ; but we may well feel sur-
prise, that these few elementary sounds can be so
variously combined and modulated as to make the
several thousand different languages which compose
human speech. The analysis of all articulate sounds
into a few elements termed an " alphabet " was accomplished at a time reaching back into fabulous anti-
quity ; but succeeding ages have been unable to much
improve the original great achievement. That different languages should contain similar words is a
natural consequence of the foregoing considerations,
without the common assumption of a communication
between the several nationalities ; and even an occa-
sional similarity of meaning, in different countries,
among words similar in sound, is within the compass
of ordinary casualty, — all men being organized alike
physically, emotionally, and intellectually ; and hence
what induced one _people to apply a given name to

any thing may have induced another people. Whether language is a human invention or an inculcation of God is often mooted ; but God made our hands, feet, heart, and lungs so cunningly as to develop their own capacity, — and why not our organs of speech also ?

As alphabets teach us indirectly the capacity of our organs of speech, so grammars teach us also indirectly the capacity of our intellect. By classifying all words according to the respective ideas which words signify, only, at most, ten essentially different kinds of ideas are found. They are called " parts of speech," and ordinarily named " articles, nouns, pronouns, verbs, participles, adjectives, adverbs, prepositions, conjunctions, and interjections." In neither grammars nor alphabets is the analysis merely conventional; but the parts of speech are founded in the organism of the intellect, and the letters of the alphabet in our vocal organism : hence the grammars of all languages are essentially alike, as are the alphabets. Indeed, grammars are more alike than alphabets, — some languages containing a few sounds that are not of universal utterance, while every language contains the same parts of speech, though words are not always classified alike therein by grammarians of even the same language ; some grammarians making as few classes of words as they deem compatible with the inculcation of syntax, while other grammarians make as many classes as they can find classifiable

differences in our ideas. But no ingenuity can invent an articulate sound that alphabets cannot spell,
nor can it invent a significant idea that the parts of
speech cannot classify; and so definitely, that, the
first time we meet in a sentence any new word, its
meaning will suggest the part of speech to which it
belongs, just as the structure of a newly discovered
plant will suggest the botanical class to which it
belongs.

That all ideas are susceptible of such a classification must be consequent upon an organic inability of
the intellect to think ideas out of the range of the
parts of speech; or else the limitation is consequent
upon an absence, in the universe, of any subject out
of the range of the said parts of speech. Either
alternative leaves the important and pregnant fact,
that, how discursive soever may be verbal speculations and contemplations, our intellect is restrained
to a given circle of ideas, just as our lungs can inhale
only certain gases, our stomachs digest only given
substances, and our hands clutch only objects of a
given consistency.

Our intellect is limited not only to the ideas which
compose the parts of speech, but the ideas are severally capable of sustaining only definite characters:
as, for instance, nouns sustain a difference of gender,
number, and case; adjectives, of positive, compara-
tive, and superlative; and each of the remaining
classes of ideas possesses well-defined characters, in

which alone they can act : while the rules of syntax
are the cycles in which alone our ideas can intelligi-
bly revolve ; for, even when syntax is violated, the
violation affects not substantially the sequences of
ideas to which the rules refer. Grammars, therefore, exhibit the intellect turned inside out, so far at
least as ideas can be expressed in words ; but the
intellect will not be satisfied with this outside inspec-
tion of its inner mysteries, and will fain discover why
it is thus limited, and especially what constitutes its
own personality : but, alas ! we possess no means for
such introspection, except by evolving outwardly
some more cycles of the intellect, of the same kind
as the former. If these will satisfy a man, few persons are so dull as to be unable to make, on the
subject, some theory satisfactory at least to his own
intellect, and possibly satisfactory to his own alone.

III.

THE PREDESTINATE IDEAS OF THE INTELLECT.

WHEN a lighted match is applied to gunpowder, and a flash ensues, with an explosion, every person present sees the flash and hears the explosion; his senses requiring therefor no preliminary instruction : and simultaneously his intellect will, without any previous instruction, conceive power and causation in the match, and effectuation in the flash and explosion; the intellect conceiving thus by its instinct, as the eyes by their instinct see the flash, and the ears hear the explosion. Accident may regulate what shall transpire momentarily within the purview of a man's senses; but how each sense shall be affected thereby is not accidental, but predestined by the organism of each sense respectively. Hence men converse together about their common sensible experience, with confidence that the senses of all men respond identically. Accident regulates, in the same degree, what shall transpire within the purview of a man's intellect : but how his intellect shall be affected thereby is predestined by the organism of his intellect, just as the organism of every vegetable predestines what it shall extract from the earth ; the tobacco plant extracting only what produces tobacco, and the plant of corn extracting what produces corn.

The above recital manifests the existence of two different principles in the acquisition of knowledge : a principle of casualty, which determines what shall momentarily come to pass within the purview of a man's senses and intellect ; and a principle of fixation, which determines what a man's sensible and intellectual organisms shall respectively respond to what comes to pass. A man may every day, during a long life, see some object — an elephant, a comet, a plant — which he never saw before ; but the visibility which is new is only some new arrangement or modification of color, light, and shade, which are fixed, predestinate impressions produced on every man's eyes by all visible objects. Without intending to insist that sight can take cognizance of only color, light, and shade, I adopt the definition because it answers my design of discriminating between the knowledge which is organic in the sense of sight, and the knowledge which we acquire casually of objects that are visible. The like may be repeated of the intellect : its predestinate ideas are few, though the number and variety of their applications seem illimitable. But we have not analyzed into their elements the predestinate ideas of the intellect, as we have the predestinate perceptions of the eyes into color, light, and shade. What ideas, then, are predestinate, and which the intellect must conceive, and to which its organism is restricted? A preceding essay has shown that they are articles, nouns, pronouns, adjectives, verbs, participles, adverbs, conjunctions, prepositions, and interjections; but this classification makes no distinction between nouns, &c., that are predestinate, and other nouns, &c., that are not predestinate. As, for instance, let us assume that the noun "power" is one of the predestinate ideas which the intellect must conceive, whenever any thing capable of exciting it comes within the purview of the intellect: yet, after the idea of power has been thus acquired by the intellect, it need not know that cold has power to solidify water, and heat to evaporate it; that fire has power to explode gunpowder, and gunpowder to rend a rock, &c. These, and innumerable other manifestations of power, are not predestinate ideas; but every man is continually learning them, casually or otherwise; and the manifestations thus learned may be different in different men: while power itself has ever been known to all men alike; and the same may be said of every other of the intellect's predestinate ideas. They are not innate, so as to develop themselves spontaneously; but every part of the earth possesses excitants to the conception of all the predestinate ideas: and thus is preserved the identity of man's intellectual knowledge, so far as relates to its kind; just as the identity of man's muscular performances is preserved through all periods by the identity which exists in man's muscular organism, and by a universality of the excitants which call the muscles into action. The predestinate ideas of the intellect are discriminable, therefore, from the casual or experimental ideas, as the fundamental rules of algebra are discriminable from the equations and problems which the rules evolve, and as the predestinate movements of man's hands and fingers are discriminable from the paintings and sculptures which the predestinate movements evolve.

But, after we admit that the intellect's conceptions are limited in kind to certain fundamental or predestinate ideas, the limit may be consequent to either the organism of the intellect, or to an absence in the surrounding universe of any things except what the intellect can take cognizance of. The practical result will be alike under either alternative; and we possess no means of deciding between them. Still, the fact is interesting, that how much soever physical experience, intellectual reflections, and emotional manifestations, may increase our intellectual knowledge, it must all be composed of a limited number of different predestinate ingredients: just as our houses, how much soever varied in form or structure, must all be composed of materials whose variety of kinds is predestinate and few; our industry being able usually to increase the quantity, but not the kinds. What then, in detail, are the predestinate ideas to which our intellect is restricted? My design includes no settlement of the question, but to assert the less debatable proposition, — that our intellectual speculations are like a game of chess, wherein the pieces are limited in number and definite in character, while the mode of playing them may be infinite. The wisest philosopher, therefore, like the most skilful chess-player, must still employ the same instruments ; and he will differ from less wise persons in only the use he will make of the common instruments : as, for instance, the *cause* which sustains the world is no longer the shoulders of Atlas ; but the intellect, constrained as originally by its organism to impute a *cause*, has conceived, from its present knowledge of casually acquired causes, a better cause than Atlas.   Light and darkness are no longer deemed *effects* of a revolution of the sun around the earth ; but the intellect, organically constrained, as originally, to deem them *effects*, has, from its present knowledge of casually acquired efficients, selected a better efficient.   Earthquakes are no longer *connected* with the struggles of imprisoned Titans ; but the intellect, organically constrained, as in all past time, to *connect* them with some precedent, has frequently changed the precedent as its store of casually acquired knowledge supplied one that was more satisfactory than the others.   Problems like the foregoing, in which power, causation, effectuation, time, place, resemblance, number, quality, relation, connection, and the intellect's other organically predestinate ideas, are the staple, have ever engrossed men's speculations ;

though new answers are continually superseding old ones, by reason that man's acquired knowledge is ever supplying new analogies : so that every generation of men have smiled at the simplicity of the past, and some future generation will smile at the best speculations of the present; and so onwards, world without end.

This view of intellectual speculations was taken by Aristotle, whose categories were an attempt to analyze into a few predestinate ideas all that the intellect can conceive ; and now, after the lapse of twenty-three hundred years, the chessmen (to continue the metaphor) remain unincreased in number and unchanged in character. We may not agree with Aristotle in the number of the intellect's predestinate ideas, or in the naming of them ; and, twenty-three hundred years hence, these questions will be as unsettled as they are at present; though in relation to a limit in the kind of our ideas, and that the varieties are few, no difference can at any time exist among reflective men who will take the trouble of analyzing into specifically different ideas any mass of intellectual speculations. We often complain that novels are but a new adjustment of old topics; but we may make the same complaint of landscapes, when we travel : we can see only new combinations, and but rarely new components. We may make the same complaint of our food : the bills of fare are only repetitions of the same materials differently prepared.

In short, novelty of ingredients, in any department of art or nature, is enjoyed by only the young; and a man need not attain any great age, before he may say with Solomon, " The thing that hath been is that which shall be, and that which is done is that which shall be done; and there is no new thing under the sun," — nothing new, because created ingredients are either limited in variety, or our organisms, intellectual, physical, and emotional, are limited in their apprehension.

IV.

THE RELATION OF OUR THREE ORGANISMS TO EACH OTHER.

MOST animals, certainly all the most perfect in forma-
tion, possess a physical personality, an emotional
personality, and a quasi-intellectual personality, and
are thus essentially triune. The physical personality
includes the corporal senses, which, with other corporal organs, evince an adaptation to the external
universe replete with suggestive interest; but the
adaptation has been often shown, and my theme is
only the relation of the three personalities to each
other. The emotional and intellectual personalities
of every animal seem subsidiary to the physical personality and complements thereof; for they gradually
subside on the subsidence of the physical powers,
whose total extinction by death always occasions the
evanition of both the other organisms; though occasionally the physical powers outlive the intellect and
the emotions.

The Relation of the Emotional Organism to the Physical.

The relation of the emotional organism to the
physical is seen in the emotional change produced

in ·all male animals by physical emasculation. Animals, also, that generate without physical intercourse with others of their kind, are devoid of various emotions discoverable in copulatory animals, and useful to their copulatory functions ; just as web-footed birds desire immersion in water, and claw-footed birds fear immersion. Longevity diminishes the physical powers of man, and impairs in an equal degree his correlative irascibility, ambition, restlessness, and impatience ; but, wealth being more necessary to the feeble than the strong, avarice is rather increased than impaired by physical weakness. So subservient, indeed, to our physical requirements, are our emotional cravings, that they are deemed by some physicians remedial indications : hence cold water and cold air are no longer withheld from fevered patients, and inappetency is indulged on the same principle. Contrast also the emotions accompanying different physical formations. Man requires society ; and so controlling thereto are his emotions, that solitary confinement is the most cruel of inflictions : but beasts of prey are best sustained by roaming singly, and they are emotionally solitary. Emotional fear is as universal in animals as physical danger : but emotional courage is restricted to animals that possess powers of physical resistance, and usually is proportioned thereto ; though, when fear cannot secure an animal's safety by retreat, courage rises in even the most timid, as a last remedy against unavoidable danger. Analogous to this is the courage which hunger develops, and which augments with increasing necessity, and diminishes with satiety; so that lions and tigers are occasionally pacific, and timid animals ferocious. The rhinoceros possesses enormous strength and aggressive resources; but, incased within an almost impenetrable skin, it needs no defensive courage, and is, accordingly, destitute thereof; while a wasp, exposed all over to destruction, is so proverbially irritable, that contact with its petty weapon is carefully avoided.

The Relation of the Intellect to the Physical Organism.

That man's physical organism is worthless without an intellect, is seen experimentally in idiots; and a like defect would produce a like result in all animals. In matters within their physical power, they all evince intellectual capacity; but, in matters beyond their physical powers, all animals seem equally incapacitated intellectually. A spider, when pursued, will roll his body into the form of a ball, and thus elude detection by enemies not acquainted with the trick; and, in playing with a captured mouse, a cat will avert her eyes therefrom to encourage an attempt to escape, whereby to enjoy a recapture. Birds allure a pursuer from their nests and their young by alighting at some distance therefrom. A coach-horse will walk to his stable when he becomes unharnessed,

and, at the proper time, will retake his position before the coach ; but, the adjustment of his harness
being beyond his physical powers, it is to him an
inextricable mystery, as the periodical migration of
birds and fish are mysterious to us. From some rat-
traps an exit would be easy to an animal organized
to lift a fallen door ; but, should the confined rat see
you lift the door, the escape thitherward will be
obvious to him : but how to cause such a means will
remain as inscrutable to his intellect as it is impracticable to his physical powers ; just as a man confined
within four insurmountable stone walls may see them
suddenly opened by an earthquake, and the means of
escape become apparent to him ; but, the production
of an earthquake being beyond his physical powers,
its production is an inexplicable mystery to his intellect.   When our Saviour says to the impotent man,
" Take up thy bed, and walk," the intellect conceives
curative efficiency in the mandate ; but, a recovery by
mandate being beyond man's physical powers, the
effect of the Saviour's mandate is mysterious. I once
saw a farce enacted, entitled " Harlequin Miller."
Old men and women were placed in a mill, and,
after being mysteriously ground, came out young
and blooming.   Gross as is the analogy on which the
farce is founded, the intellect can conceive no mode
of rejuvenation but what will be equally monstrous ;
thus manifesting the impotence of the intellect, except
in subservency to man's physical powers.   When a

man's leg becomes broken or dislocated, the means of relief, being within his physical powers, are within the purview of his intellect: but, when a horse breaks one of his legs, no reparation is within his physical capabilities, except to refrain from resting his limb on the earth; and his intellect conceives that remedy alone. A dog, when wounded, is incapable of any physical appliance, except with his tongue to lick the wound; and his intellect suggests this process, and can conceive no better. We sometimes think revelation might properly have gratified our curiosity by informing us how the creation of light followed the fiat, "Let there be light;" how "God formed man of the dust of the ground;" and how "the rib, which the Lord God had taken from man, made he a woman:" but, unless the modes were within our physical capabilities, — as we know they are not, — our intellect could not have understood the information, had it been attempted. By analogizing the unknowable to something within our physical powers, we learn all we are capable of knowing; and this is accomplished in the foregoing cases by the above narratives: we being accustomed to fashion pottery out of clay, as, we are told, Adam was made; and to fashion images out of bone and other materials, as, we are told, Eve was formed; and accustomed "to say to this man, Go, and he goeth; and to another, Come, and he cometh," — as, we are told, light was created. Every biblical miracle is made intelligible to us by means like the foregoing ; as, when our Saviour opened the eyes of the blind, " he spat on the ground, and made clay of the spittle, and he anointed the eyes of the blind man with the clay."

That our physical powers limit our intellect's comprehension, may be further manifested by our intellect's inability to conceive that two apples added to two other apples make either more or less than four apples. The inability is only a consequence of the physical inability, and hence is not apparent to a child or man till he knows experimentally or analogously the physical inability. Why, on the contrary, can the intellect conceive that one drop of quicksilver added to another drop will make only one drop? Because such a result is physically attainable. Why cannot the intellect conceive that a billiard-ball can be both very hot and very cold at the same time? Because we find the combination physically impracticable. But why can the intellect conceive that the same ball can be, at the same time, very hot and very black? Because we find that such a result is physically attainable.

The Relation of the Intellect to the Emotional Organism.

But the intellect is subsidiary to the emotional organism as well as to the physical. Our amusive literature, and our theories, hypotheses, conjectures, guesses, deductions, and inductions, are responses of our intellect to our emotional curiosity, inquisitiveness, expectations, hopes, fears, doubts, beliefs, and unbeliefs, &c. Theological doctrines are responses of the intellect to our emotional piety, superstition, reverence, fear, hope, &c.; and hence theology is as old as man. Rage and revenge have always created some intellectual system of warfare; curiosity has always supplied an astronomy and cosmogony; the desire of life, and the fear of death, have always originated some hygiene and therapeutics; love and jealousy, some system of marriage to appropriate particular women to particular men: just as dams have always been constructed by beavers, nests been built by. birds, and burrows made by rabbits.

But, while our intellect is organized to respond thus to our emotions, its responses are changeful; they being limited at every period to the physical knowledge of the period, — the intellect knowing only such agencies as it acquires through the external senses: therefore, before the invention of gunpowder, the intellect could not account for earthquakes on any theory of explosion; before the discovery of steam-power, it could not account for them on any theory of steam-expansion; before the , discovery of magnetism, it could not account for tides on any theory of lunar attraction; before the discovery of electricity, it could not account for the aurora borealis and lightning on any theory of electricity,

&c. The sun, which is first seen in the east, disappears in the west : and, when curiosity seeks an explanation thereof, the intellect can give only two intelligible responses, — either the sun moves round the earth from east to west, or the earth around the sun from west to east; and simply because our physical organism has supplied the intellect with no other analogous *modus operandi*. Possibly at some future time, and by some physical experience now unknown, the above explanations may be deemed as childish as the fabled theory of old, that the earth is sustained on the shoulders of Atlas, — a fable only noteworthy as manifesting the dependence of our intellect on the physical knowledge of the period. "How are the dead raised up ? and with what body do they come ? " St. Paul's intellect answered these questions in the only way they can ever be answered, — by some physical experience analogous to the questions : " Thou fool ! that which thou sowest is not quickened, except it die : and thou sowest not the body that shall be, but bare grain ; it may chance of wheat, or some other grain. But God giveth it a body as it hath pleased him, and to every seed his own body." The more unique, therefore, the subject is that our curiosity would investigate, the more unanalogous thereto must be our sensible experience, and ratably anomalous and wonderful must be the theoretical solutions thereof by our intellect : hence some of the startling speculations of astronomy may

possibly originate in a want of analogy between terrestrial and celestial objects and operations; and may
we not reasonably suspect such a diversity, when, by
assuming the contrary, we prove mathematically that
the nearest fixed star is so distant, that a cannon-ball
projected therefrom with the greatest known velocity
of such a projectile, and moving towards the earth,
would not reach it in less time than seven millions of
years ; and that suns probably exist, whose light has
been travelling towards the earth many millions of
miles every minute since the creation of the universe,
and it has not yet arrived ?   Subject to a like defect,
all the intellect's solutions of vital operations, generation, sustentation, disease, growth, &c., have been but
little improved since the earliest ages, and they never
can improve essentially : therefore physicians no
longer elaborate remedial theories from physical analogies, but rely for success on analogies from success-
ful remedial experiences.

The Two different Organic Powers of the Intellect.

Our intellect possesses, therefore, two distinctly
different powers, which have not been heretofore
discriminated, — a power responsive to our physical
organism, and on which is founded all our arts ; and
a power responsive to our emotional organism, and on
which is founded all our speculations.   Providence
would hardly have thus organized us, if the responses

in either branch were merely illusory; though our
intellect (which includes our judgment) often with-
holds its full approval of many of the intellect's
responses to the emotions; while, on the contrary,
our emotional organism (which includes the feelings
of belief and unbelief) yields belief equally to both
classes of the intellect's responses: hence the expectations of Rationalists will never be realized, that a
period will arrive when nothing shall be believed,
unsanctioned by the judgment. Speculations change
their form; but speculation as a process is as per-
manent as sensible knowledge. We shall, therefore,
always possess and believe a theology revealing to
us the everlasting future, a geology revealing to us
the pre-Adamitic past, and an astronomy which includes more wonders of the present time than theology and geology yield of the future and the past:
still, theology, geology, astronomy, and all other
speculations, must be fashioned at any given time
by our then physical knowledge; and thereout our
speculations cannot wander, any more than our feet
can wander beyond the boundaries of the earth.

.

V.

THE POLARITY OF INTELLECTIONS.

No blind person can recognize either darkness or
blackness, though both are ever present to him ; nor
can the deaf recognize silence. Hence recognition
requires more than the presence of recognizable
things ; and this requirement the following postulate
supplies : namely, every intellection includes the recognition of a negative as well as a positive ; and the
two poles must concur to produce in any person an
intellection. Darkness cannot be recognized by the
blind ; they knowing only the positive, but not
the negative, — not darkness : and, if light were as
unintermitted to us as darkness is to the blind, light
would be a positive without a negative, and therefore it would be as unrecognized by us as darkness
is by the blind. We should enjoy all the benefits
of light as we now enjoy them ; but we could not
recognize it, by reason of our not knowing what
not light is, — just as the blind suffer all the in-
conveniences of darkness, but they cannot recognize
it. If death were unknown, and immortality were
the condition of all things, life would be a positive
without a negative, and therefore unknown to us :

4

just as a statue is undiscoverable in a block of marble, till all the marble is cut from the statue, so as to make a negative thereto; and just as a man's face is undiscoverable on a painted canvas, till all but the face is negatived by being colored differently from the face; and, the more we contrast the positive and negative descriptions, the more distinct becomes the face. A sheet of white paper contains various profiles of man and animals; but they are severally indiscoverable till we make a negative to each positive by cutting from the paper (or manifesting otherways) the parts of the paper not included in the respective profiles. If the atmosphere was not occasionally agitated into winds, a man might live to old age without recognizing the presence of air. The facility we experience in detecting fish in water, or a bird or an insect amid foliage, is proportioned to the contrast in color that exists between the object and its surroundings. Writing is legible in proportion as the written and unwritten parts negative each other: ink and writing-paper are good in proportion as they best accomplish these purposes. Darkness produces invisibility only as it prevents any negative to the enveloped positives; but snow can be recognized almost as distinctly in darkness as in light; both being equally a negative to the appearance of snow. Darkness increases the visibility of the moon; for darkness, better than light, negatives all that is not moon. Could we see the sun, with the surrounding light negatived, as we see the moon in a dark night, the sun's positivity would be enhanced ratably with the negativity. A smoked glass produces the result partially; but, while it negatives the sun's surroundings, it impairs the sun's brightness. The sun shining on fire and flame seems to deaden them, but only by impairing the surrounding negative. An opposite effect is produced on the furniture of a room; for, when shined on by the sun, it will often appear covered with dust that was invisible previously, from the absence of a negative thereto. A fog produces obscuration by obliterating all negatives to the objects it envelops; but this is sometimes partially obviated by an increased blackness with which some objects tinge an enveloping fog, and to which the untinged fog becomes a negative. A surrounding daylight is as obscuring to the moon as a surrounding fog; both yielding an equally imperfect negative to the moon. A telescope transforms the galaxy into separate stars, not by merely magnifying them, as is usually alleged (for that alone would not effect the intent), but by surrounding each with a negative. The common spectacles that are worn to correct impaired vision act on the same principle; increasing the negative surroundings of an object, rather than magnifying the size of the object.

The omnipresence of God must remain unperceived by us, though it may possess more positivity around us than a noonday sun; the premised omnipresence permitting no negative of the positive presence. The revolutions of the earth are unfelt by us, how violent soever may be the gyrations; no negatives making the positives perceptible. If we can suppose a man to be on shipboard from his birth, and the ship to be unceasingly rocked with any unvariable motion, he would be unable to recognize the motion : nor could it be manifested to him by any instruction from other men differently conditioned, any more than a blind man can be instructed to recognize darkness; the instruction requiring for its removal the knowledge of a negative. The pressure of the atmosphere is unfelt on our body; the pressure being uniform and unremitting, a positive without a negative. Fish can have no consciousness of being wet; as our constant immersion in air renders us unconscious of any effect therefrom on our skin. When, however, our cuticle is wounded, the air produces thereon a feeling evincive that our former insensibility was owing to only the universality of the air's action on us and the absence of a negative. How many positives exist around us and within us, that are unrecognizable by reason of their unremitted presence and the consequent absence of a negative, we possess no means of conjecturing; but they may be numerous.

The polarity of intellections is indicated in every language by a responsive verbal negative to almost every affirmed positive; as, light and darkness, sapidity and insapidity, colored and uncolored, above and below, in and out, hot and cold, light and heavy, high and low, soft and hard, long and short, easy and uneasy, rough and smooth, straight and crooked, handsome and ugly, &c., &c.; and in each case the positive would be unrecognizable without the negative. Hence the common paradoxes, that without pain there would be no pleasure; without disease, no health; without cowardice, no courage; without poverty, no riches; without pride, no humility; without vice, no virtue; and, without the existence of an exception, we should have no rule, &c.

Some positives seem to conflict with the foregoing doctrine. All visible objects are said to be colored; hence color is a positive without a negative: and this should make color unrecognizable, and the word "uncolored" meaningless. But "color" is used in a sense that makes its universality untrue, white being a negative of color; and other negatives exist, as water, and the glair of eggs. Figure and shape are said to be attributes of every visible object; and therefore, being positives without a negative, they should be unrecognizable. The premises are, however, untrue, as is evinced by the words "shapeless" and "formless," which denote practical negatives to figure and shape. Philosophers have long been divided in opinion as to the existence of a vacuum; some affirming that the universe is a plenum: but a plenum is significant only as the negative of a vacuum, and *vice versâ;* hence, without both, neither is significant. The like may be said of externality; which some philosophers deny, insisting that all we know is internal of our mind: but externality and internality are reciprocal conditions, and neither is significant without the other.

But if we can recognize no positive till we know a negative thereto, and *vice versâ,* how can we commence the acquisition of knowledge? Nature has provided for the occasion; the normal condition of each sense constituting a negative thereto: hence his saliva is the natural negative by which an infant will recognize a taste in the first drop of his food, and the normal temperature of his body is the negative by which he will recognize heat and cold. A child born in a prison, surrounded by fetor, will, by the foregoing principle, be unconscious of the odor; for, while any fetor continues uniform and incessant, it will constitute his olfactory negative. A man's normal state of health constitutes his negative of pain; and the negative differs in different men: well-formed men differ therein from men who are congenitively blind, deaf, humpbacked, or deformed in any other manner.

The foregoing examples may never occur in the degree assumed: but the principle they enunciate is true; and any man's normal negative may change temporarily or permanently. The operatives in a factory, where the clang of machinery is uniform and incessant, and the inhabitants of a populous city, and the mother of a group of clamorous children, soon deem the accustomed sounds a negative of noise, while it would be distracting to persons unaccustomed thereto. Men can so familiarize themselves thereto, that the taste of tobacco may become a negative; furnacemen may come, by habit, to deem as a negative a degree of heat insufferable to persons unused thereto; and persons long diseased become unconscious of pains that originally were very afflictive. When green spectacles are first worn by any person, all objects seen therethrough are tinged with green; but, after a certain persistency in such glasses, their color becomes a negative, by which other colors are perceived, the same as if the spectacles were made of uncolored glass. Men afflicted with strabismus possess a negative condition of sight that differs from the negative of other men; but all men recognize objects equally well, though with different negatives.

On the foregoing principles, every man's locality is his negative of distance; and his muscular strength, the negative of weakness: all persons are young who are younger than himself, and all old who are older. By a like standard, he determines who is rich, and who poor; who is short, and who tall; who is fat, and who lean; who is quick in motion and intellect, and who slow; who is witty, and who dull. What promotes his health is healthful; what is prejudicial to him is hurtful; what administers to his wants is useful; what is not serviceable to him is

useless; what man cannot bound, he deems bound-
less; what he cannot fathom, becomes unfathomable;
what he cannot understand, becomes incomprehensi-
ble; and what he cannot enumerate, becomes innumerable. In relation to this particular positive, some
Indian nations cannot number beyond the aggregate
of their fingers; and then innumerable commences.
This probably exaggerated example shows how men
differ in estimating qualities; but men's predilections
and capacities are enough alike to obviate much mis-
understanding in our intercourse with each other.
Still, the better we understand the criterion by which
individuals form their judgments, the less we shall be
irritated by deviations from uniformity.

I cannot satisfactorily say more on the polarity of
intellections; though I see indistinctly that the assumed polarity, perhaps a bi-polarity, underlies the
comparison of adjectives; underlies also the meta-
physical necessity, that "a whole must ever be greater
than a part;" and underlies the intellect's tendency
to reject the "golden mean," and to vibrate in medi-
cal treatment from heating to cooling, from exhausting
to stimulating, from allopathy to homœopathy, and to
vibrate in criminal legislation from the execution
of criminals with the utmost publicity to their execution with the utmost privacy, from the infliction
of death for slight offences to the abolishment of capi-
tal punishments for the highest, from democracies in
government to despotisms, and from despotisms to

licentiousness; and underlies all disputation, every assertion exciting in the person addressed a negative response, and every negative exciting a positive; and underlies all the antithesis of literature, as in the Bible, "I am Alpha and Omega, the beginning and the end, the first and the last;" and in Shakspeare,—

"Be thou a spirit of health, or goblin damned;
Bring with thee airs from heaven, or blasts from hell;
Be thy intents wicked or charitable;" &c., &c.

I will close the topic of polarity by adverting to eternity, infinity, omnipotence, &c.; of which we usually say, that their full meaning is unknown to us. And why is it? Because they are severally only negatives. We, for instance, know time as a positive, but eternity (time without end) we know as only a negative of time with an end; we know space as a positive, but infinity (space without bounds) we know as only a negative of space with bounds; we know power as a positive, but omnipotence (power without limitation) we know as only a negative of power with limits. Our knowledge of death is like the foregoing; it consisting mainly of negatives. The same may be said of sleep. Nor can we find out God to perfection, says the Bible : and the organic obstruction consists in polarity, which limits our knowledge of God to negatives, as in the above example of his omnipotence; the like obstruction applying to all his attributes.

VI.

WHAT WE THINK, HOW WE THINK, WHAT WE CANNOT THINK, AND THE ORDER IN WHICH WE THINK.

CHAPTER I.

WHAT WE THINK; OR, THE ANALYSIS OF THOUGHTS.

THOUGHTS are divisible into six kinds, as follows : —

CLASS 1. —*Visual Thoughts.*

If I think of the moon as a sight, it is a visual thought. When Hamlet exclaims that he can see his father " in his mind's eye," the exclamation refers to a visual thought. We can think thus of much that we ever saw; and such thoughts will often be almost as vivid as actual vision, others will be indistinct, but each will endure only momentarily. Many persons court sleep by thinking visually of sheep, and counting them as they leap successively over a gate from one field into another.

CLASS 2. —*Verbal Thoughts.*

We can talk in thought as distinctly as we talk audibly. When the Bible commands us " to com-

mune with our hearts upon our bed, and be silent,"
we recognize the verbal thinking to which the pre-
cept refers.   Hannah spake thus " in her heart : her
lips moved, but her voice was not heard."   An
accountant can reckon in verbal thought ; a mathe-
matician can demonstrate therein a problem ; an
orator can declaim in the same way ; and a Christian
can thus pray, or repeat the Decalogue.   A Frenchman thinks French words ; a vulgar man, vulgar
words ; an uneducated man thinks solecisms ; and
a profane man thinks blasphemy.   Indeed, verbal
thinking needs for its recognition only remarks
enough to manifest the process referred to.

Class 3. — *Auricular Thoughts.*

We can think of thunder as a word, and it will be
a verbal thought ; or we can think of it as a sound,
and it will be an auricular thought.   When we listen
to conversation in a language unknown to us, the
intonation constitutes what we chiefly recall in
thought.   A musician can recall in auricular thoughts
the sounds of his accustomed instrument, and the
sounds will be as distinct to him as in actual hearing.   Written music, such a person reads in auricular
thoughts as he reads written words in verbal thoughts ;
a violinist reads the written music as sounds of a
violin ; an organist, as sounds of an organ ; just as
Arabic numerals are read by a Frenchman in French
words, and by an Englishman in English.   The ability

to read in auricular thoughts the written notes that he sees, enables a musician to judge thereby the melody of the tune, about as well as he can by hearing when the tune is played instrumentally. Musicians occupy themselves with music in auricular thoughts as importunately as other men occupy themselves with words in verbal thoughts. Auricular thoughts are, of course, absent in persons void of the sense of hearing, as verbal thoughts are absent in a mute who has not learned to articulate words. Verbal thoughts are absent also in infants before they acquire speech, though they possess auricular thoughts. Bell-ringers and chimers think vividly the accustomed sounds ; and, in mania, the sufferer, in some cases, mistakes his auricular thoughts for audibly uttered sounds and words. Indeed, every person occasionally mistakes for the sound of a bell or a clock, &c., some auricular thought thereof which is excited in him by surrounding circumstances.

CLASS 4. — *Sapid Thoughts.*

When we recall in thought the taste of sugar, we are thinking a sapid thought. Every time a deaf mute thinks of sugar, he probably thinks of it in sapid thought. The emotional feelings of disgust or delight that accompany any taste, accompany usually, in a modified manner, our thought of the taste : hence conversation that excites the recollection of disgustful tastes is avoided by delicate people, especially during periods of eating. In some men, you can at any time excite nausea by speaking to them of Glauber salts. Many persons experience a flow of saliva by thinking of lemon-juice ; and a like effect is often produced by the recollection of some food of which we are particularly fond : and from this arises the expression of making a hearer's " mouth water " by some narrative of delicacies. Phraseology daguerreotypes thus our organic powers, and constitutes a common - sense philosophy, that merits more consideration than it has received. We recall more readily, in sapid thought, the taste of sugar, salt, and vinegar, than we can insipid tastes. Much, however, depends on cultivation : connoisseurs of wines taste distinctions therein unobserved by ordinary drinkers, and recall them vividly in sapid thoughts. The like may be said of connoisseurs of tea ; but, generally, men so habituate themselves to verbal thoughts, that the other classes are comparatively indistinct from unfrequency of use. A man who is deficient of any sense becomes more than ordinarily acute in the senses which he possesses, and their sensations are recalled in thought with a vividness ratably unusual.

CLASS 5. — *Nasal Thoughts.*

The odor of a rose most persons can readily recall in thought ; and I term it a " nasal thought." Laura Bridgman is deficient of hearing, seeing, and tasting ;

but her sense of smell remedies remarkably these
deficiencies, her nasal thoughts possessing great vividness. A person who visits a hospital while he is
under fear of any contagion, will think so vividly, in
nasal thoughts, of any odors he encountered at the
hospital, as to complain that he cannot rid himself
thereof; and hence the employment, in such cases, of
camphor, by which the unwelcome nasal thought is
superseded. From this organic consequence probably arises the notion, that the odor of camphor is
antidotal. That dogs and cats recall in thought
their sensations, we can hardly doubt, from the
uneasiness they often exhibit while asleep. Hounds
who follow game by a scent undiscoverable to men,
experience, probably, nasal thoughts with a vividness
unknown to us.

Class 6. — *Tactile Thoughts.*

The foregoing classific names indicate the analysis
which thoughts are susceptible of : so, when sensations of feeling are recollected, I name them " tactile
thoughts." A sufferer from gout, toothache, or any
pain, will often think thereof in tactile thoughts, not
in words. Such thoughts partake moderately of the
physical painfulness of the original feelings : consequently, topics that will excite the thoughts are
avoided by considerate companions of the sick. A
subsided toothache may be reproduced in a man by
an inquiry as to his present feelings. Thoughts of

every class are vivid in proportion to the acuteness of the original sensation. To smooth with our hand a piece of velvet, or to pass our hand over a rasp, produces feelings which we can recall with much vividness; and the like may be said of any peculiar feeling.

CHAPTER II.

HOW WE THINK.

EVERY thought possesses the characteristic of the sense or organ to which the thought pertains.

§ 1. — *Of Verbal Thoughts.*

" Our Father who art in heaven " must be uttered in thought as consecutively as in oral speech ; and all verbal thinking is equally consecutive. Indeed, verbal thoughts are words, spoken inaudibly ; and in most cases we can detect a movement of the tongue or the breath, as any person can discover if he will pronounce in thought the alphabet. When a man moves his organs of speech forcibly while thinking words, we say he is talking to himself. Men influenced by any strong passion often articulate their thoughts, and think aloud, especially when secluded from observation : but the tendency is difficult to resist, even in public ; and hence proceed the profane colloquies with which enraged people encounter each other, and which they usually regret on the subsidence of the emotions that caused the outbreak. The Emperor of Germany's ambassador endeavored to ingratiate himself with the King of Prussia by speaking disrespectfully of George the First of England, with whom Prussia was at enmity; but the Queen of Prussia, a daughter of George, was present at the conversation, and immediately exclaimed that " none but a scoundrel would speak thus of a sovereign person." This was her thought, and anger caused her to think aloud; the normal vigor of the intellect being often insufficient to control, in the wisest persons, the abnormal impulse of unusual emotions. That truth is elicited by wine is a proverbial admission of the effect of emotional excitement on verbal thinking; especially as alcoholic drinks, though they greatly affect our emotional and physical organisms, leave the intellect measurably unimpaired. Early in life, verbal thinking is not performed with the slight agency of the organs of speech about which we have been speaking; but children think aloud: hence the incessant prattle of infants, and their proneness to articulate in sleep.

Verbal thinking is usually deemed the most mysterious manifestation of our intellect; but my verbal thoughts seem as evidently the production of my organs of speech as articulated words, and neither more mysterious nor less. A paralysis of the tongue impairs verbal thinking as much as it impairs articulation; but it leaves other thoughts unimpaired.

How an abscission of the tongue would affect verbal
thinking, I should like to know. Persons slow of
speech are, I suppose, ratably slow in verbal thinking ; and rapid speech is connected with a ratable
rapidity of verbal thoughts. How far stammering
affects verbal thinking, I have never ascertained ; but
the stammerer can probably think verbally without
impediment. No man can speak fluently, who adopts
the adage of thinking twice before he speaks : the
pre-occupation of his organs of speech by thinking
will prevent fluency of utterance. A fluent speaker
must think aloud. We all think fluently.

§ 2. —*Visual Thoughts*.

Sights are remote from the eye that sees them ;
and equally remote therefrom seem visual thoughts.
When the moon is recognized in visual thought, its
position in the heavens is part of the thought. The
face of an absent friend, when seen in our "mind's
eye," is seen at a distance from the eye. The mind's
eye and the eye of the body seem a modification of
the same identity. When a person endeavors to
recall any appearance, he watches his eyes for the
thought he is endeavoring to excite : he will shut
his eyes, or cast them towards vacancy, and nicta-
tions of the eye-lids will be suspended. Morphy,
who plays blindfolded several games of chess simultaneously, can see distinctly, in visual thought, the
board and the moves.

After looking persistently at any bright object, we experience, if we close our eyes, a secondary vision, that is analogous to visual thinking. The phenomenon which results from a rather painful pressure on the eye is of a similar character; also the grotesque images which occasionally appear when a person closes his eyes, and solicits the representation. Ordinary blindness may not impair our power to recall former sights, nor even an extirpation of both the eyes; but a paralysis of the organs of sight would, I believe, impair visual thinking. Experiments have never been instituted in the above directions, the analysis of thought not being sufficiently understood.

§ 3. — *Auricular Thoughts.*

We can hear the hum of fifty persons speaking together, or the mingled sound of fifty musical instruments; and we can think in auricular thought whatever we can hear. In this comprehensiveness, auricular thinking differs from verbal thinking (which is consecutive, word by word); the difference of the two cases conforming to the powers of the organs of speech and hearing. As Hamlet saw his father in his mind's eye, he could have heard his father's voice in his mind's ear, had he endeavored to recall the voice in auricular thought. All of us can thus recall thunder, or any other familiar sound; and the agency therein of the sense of hearing is so well recognized, that I have known musicians to listen, and enjoin silence, when they were endeavoring to recollect a partially forgotten tune.

§ 4. — *The remaining Three Classes of Thoughts.*

The smell of a rose, when recalled in nasal thought, is accompanied by an inhalation of air, as in the act of smelling; and, in both processes, attention is directed to the nasal organ. So sapid thoughts are often accompanied by a flow of saliva, as in actual tasting. The homely phrase, " He smacks his lips in recollection of what he has eaten," evinces, that, in thinking, we not only see with our mind's eye, hear with our mind's ear, smell with our mind's nose, but taste with our mind's palate.

When any feel is recalled in tactile thought, the thought is located where the feel was experienced. An indolent woman who arises in the morning, from her bed, with a headache, may preserve it through the day by continually thinking of it in tactile thought; while a physically active man, who should arise with such a pain, would solicit other thoughts, and the pain would subside. A surgeon told me, that, by the continued application of cotton batting to the breast of a woman who feared an incipient cancer, she became cured; though, had she nursed the pain by constant thoughts thereof, the result might have been different. The only effect of the

cotton was to induce a cessation of the injurious
tactile thoughts. Strengthening plasters, and other
topical applications, operate beneficially by the same
means. Whether a woman, who has suffered from a
cancerous breast, can recall, in tactile thought, the
cancerous pains after an excision of the breast, I
should doubt; but, if they can be thus recalled, their
location will seem to be where she formerly felt the
pains.

The information of our several senses can be recollected by different persons with different degrees
of facility. A musician may recollect sounds in
auricular thoughts with more facility than he can
recall any other species of his sensible knowledge,
while a painter may recall sights more readily than
sounds. Verbal thoughts are the only class that we
systematically cultivate the recollection of; and this
cultivation commences usually at childhood, and continues to the end of our college course. Whether
we could not beneficially cultivate the recollection
of other classes of thought, may be worth an ex-
periment: geometry cultivates, to some extent, visual
thinking. My memory is more tenacious of sights
than of any other part of my knowledge; and much
that I think of absent persons and places is in visual
thoughts. To most persons, a recollection by verbal thoughts is better than any other; and therefore,
for the purpose of subsequent communication, a traveller should, while objects are before him, utter

or think verbally of the detail that he wishes to remember, and he will subsequently recall in thought the words more readily than he can the sights. To this principle is owing the advantage of viewing objects in company with friends; the comments made to each other being subsequently recallable easily in verbal thoughts. While travelling in a stage-coach, the late Hannah Adams kept repeating, "Trunk, band - box, carpet - bag, and satchel." A person asked her if she wanted any thing. She said, no; but she was afraid of forgetting some of her baggage if she did not commit to memory the words.

. Our physical feelings are less easily recalled in thought than our other sensations; and this providential dispensation exposes us to the evil of only actual pains, except rarely, and in the mitigated form of tactile thoughts. Emotional feelings are renewable, but they are not recallable in thought; nor are emotional feelings evanescent, as all thoughts are. The same melancholy, or other emotional feeling, may pervade a man continuously for many days in succession.

CHAPTER III.

WHAT WE CANNOT THINK.

THE blind know nothing of colors, nor of any other sight, and consequently they can think no visual thoughts; and no man, possessed of vision, can think visually of any sight that is visually unknown to him : his disability therein is precisely like the disability of the blind. This limitation of visual thoughts is organic, and of course insurmountable. I once heard a gentleman refuse to look at the dead body of his friend; assigning as a reason, that he wished to think of his friend as he knew him in life, and not as he appeared when dead. Every man understands the principle alluded to. With verbal thoughts, the person could think of his deceased friend in any phrases that language can supply; but, with visual thoughts, he could think of him only as he had seen him. Still, unless a man understands the six different manifestations that are confounded under the one general name of " thought," he may be puzzled when he is told of a blind man who once discoursed accurately on light and colors. The blind man's thoughts thereon were but verbal. A traveller in new and distant countries may publish in a book all he witnessed; but a reader thereof will acquire therefrom only verbal thoughts, while the traveller's thoughts will be the sights and other revelations that were made to him by his senses.

That a man's sensible knowledge is limited to his sensible experience, constitutes the organic barrier which hides from us futurity. Verbal thoughts may overleap the barrier ; but it is not overleaped sensibly. If any person believes he can think visually what he never saw, let him endeavor to think visually of some color that he never saw, and he will find himself unable. You can see, in visual thought, an angle, a globe, a plain ; but you will be unable to thus see some shape which you never saw. The limitation imputed thus to visual thoughts pertains to all the other classes. You can think, in auricular thought, the sound of thunder, of a flute, a drum, the neighing of a horse, the bleating of a sheep ; but you are unable to think thus of some sound you never heard. To make the limitation evident, think of some sound that shall differ from any you have heard, — not in degree, but in kind, — and you will find such thinking impracticable.

But persons may believe, that by combining what our senses have informed us of, or by subtracting therefrom, we may, in thought, create a centaur. We can think verbally a centaur, and we can think visually a man's head, and, in another visual thought, a horse's body ; but we cannot form the two sights into one visual thought. A man can combine black and white in one visual thought, if he has ever seen the two colors in juxtaposition ; but, never having seen meridian daylight and midnight darkness in juxtaposition, he cannot combine the two in one visual thought. Locke says, "We can place two lengths together, and think of increased length as resulting from the union ; but, if we place together two parcels of snow, we cannot think of the aggregation as increasing the whiteness." Both results depend on the principle, that our thoughts of sensible things conform to sensible archetypes. A man cannot bite his own nose off ; nor can he think the operation, except in words.

Water and vinegar will fuse into one taste, modified by the union ; nor can the two constitute one sapid thought, except in the manner they can be tasted. On tasting brandy, a person may say he will try it mixed with water. Why not test the mixture in thought? Because we can think sapidly of only the tastes we have experienced. A perfume is usually a combination of several odors ; but, how familiar soever the separate odors may be to you, the perfume can be experienced in nasal thought only after you have scented the compound. The instruments that compose a band, you may have often heard separately : but you cannot combine their sounds in auricular thought, unless you have heard them together ; nor can you, in auricular thought, resolve the compound into its separate sounds, unless you have heard them separately. The Bourse at Paris, during " high change," resounds with a clamor which no previous acquaintance with the human voice would enable a man to think in auricular thought ; but, having heard the sound, he will thereafter be able to recall it.

Verbal thoughts differ from all the foregoing. They are not limited to sentences previously uttered : but we can think verbally whatever we can speak ; the two processes differing only in audibility. Hence we can think, in words, of fifty noonday suns blazing together in the heavens, or lying scattered on the earth ; of grottoes at the bottom of the sea, and fires in the centre of the earth ; and of any other thing that language can express. Had we the same control over our sensuous thoughts that we have over our verbal thoughts, curiosity would no longer compel us to travel ; but, wandering only in imagination, fiction and reality would glide before us in thought like the images of a phantasmagoria, and we would soon lose the power of discriminating intellectual creations from physical realities.

CHAPTER IV.

THE ORDER IN WHICH WE THINK.

THOUGHTS succeed each other in the order that the objects thought of became known to us.

§ 1. — *The Recollection of Words.*

What letter follows R, in the English alphabet? Every man will immediately name S ; but, if you ask

him what letter precedes R, he will usually be unable
to recollect Q till he has first thought of some letter
that precedes Q, and then remembers downwards to
R.   So a person, who has learned multiplication by a
table in which the multiplier is never larger than the
multiplicand, will not know the sum of nine times
eight till he has reversed it into eight times nine.
This difficulty may continue with a man during his
life; and it illustrates the order in which we think.
A child who knows by memory the alphabet will be
unable to proceed if you stop him in any part, and
thus break the continuity of his thoughts.   He will
say, you have put him out.   We remedy such a break
by repeating the letter which the child last uttered,
and thus restoring the continuity.   All prompting
proceeds on this principle.   A child, in repeating from
memory any formula, will pronounce the words rapid-
ly; for he knows, that, in a slow utterance, some extra-
neous thought may intervene, and put him out.

To learn by rote any series of words, you must
read or repeat the words till your organs of speech
become so accustomed to the series, that every word
of it excites a recollection of its successor.   Children
soon acquire experimentally this mode of committing
to memory, and also that to repeat audibly the words
which they desire to remember is more effectual than
to repeat the words in thought.   Before a child can
add two and two, he must recollect the phrase " two
and two are four," and so of every other two numbers

to the extent of his knowledge of addition; but, as
we commence a new series of addition, every time we
attain the sum of a hundred we are relieved from bur-
dening our memory beyond every binary permutation
of figures that can make the sum of a hundred. The
number of such permutations, in every possible series
of figures in a single column whose aggregate shall
not exceed the number of one hundred, is, I believe,
nine hundred. Whether a table of these permutations, to be learned abstractly, by memory, as we
learn multiplication, would assist the acquisition of
addition, I know not. Some accountants can add two
columns at once; and I have heard of three columns
being thus summed up.   But, before two columns can
be thus added, some ninety-nine hundred binary permutations must be learned by memory; while, to add
three columns, requires, I believe, ninety-nine thousand nine hundred permutations. The addition of
two columns together is, therefore, about ten times
more difficult to learn than the addition of a single
column. Most people limit their *memoriter* multiplication to twelve, both as relates to the multiplier and
the multiplicand; but no natural limit exists therein,
but the capacity of man's memory. A *memoriter* multiplication up to twenty, for both multiplier and multiplicand, is easily acquired; and to add up two
columns of figures simultaneously is not difficult.

§ 2. — *The Recollection of Sights.*

If I have frequently seen four coaches pass my window in succession, the first black, the second red, the third yellow, the fourth green, my visual recollection of the carriages will present them in the same order; and, were I called to testify in a court of justice as to the succession of the coaches, the order could be recalled in visual thought, with little chance of misplacing the sequence. Should a person, however, testify falsely of such a sequence, he would be in constant peril of forgetting the order to which he had testified; possessing no monitor thereof but a recollection of his previous words. Hence the proverb, that liars should have good memories. Witnesses are often embarrassed when counsel will not let them testify to events in the order of their occurrence. The narration desired by the witness may contain much that is irrelevant to the issue on trial; but it is necessary to his recollection of the parts which are relevant.

§ 3. — *The Recollection of Sounds.*

A musician is frequently unable to recollect a tune; but, when another person will commence it, the musician can sing the rest without prompting. The cuckoo is a bird very common in England, and it issues a cry from which the bird's name is derived, — "Cook-oo." The cry is easily imitated by the human voice; and whoever is familiar with the cry, will recollect, in auricular thought, the second syllable as soon as he hears the first : but, should any person utter only the second syllable, it will probably be unintelligible to the hearer. The like may be said of the " whip-poor-will " of America, or the " katy-did." The principle is more than the association of ideas ; the first syllable of cuckoo being as much associated with the second as the second is with the first. Still the association aids not memory, except in the sequence to which the senses are accustomed.

§ 4. — *Tastes, Feels, and Smells*

Are all recollected in the order that our senses have manifested to us the originals. The order is independent of the will, and providentially prevents our remaining absorbed in the contemplation of any single thought ; our thoughts flowing onward unceasingly, whereby any painful recollection is soon left behind till it is again recalled by some of its antecedents. A principle of succession, kindred to the foregoing, regulates our muscular recollections or repetitions. A person, after having played a tune frequently on the piano, will unpremeditatedly move his fingers on the correct keys ; but he will be as unable to move the keys in a reverse order, till practice has familiarized him to such moves, as he is to spell his name in an inverted order.

C O N C L U S I O N.

I DESIGNED to end with a summary of what the foregoing pages discuss ; but I find myself unable to
epitomize what together is little more than a very brief
epitome of several books heretofore published by me.
I therefore, in conclusion, only iterate what the title
promised, — that the contents are explorations into
the depths of intellectual knowledge by means of an
ultimate analysis thereof. Our physical possessions,
though illimitable in variety, are, in their essence,
unknowable beyond the few elements into which they
are analyzable ; and the like is true of our intellectual
possessions. Curiosity desires to delve below, soar
above, and advance beyond, these mysterious barriers ;
but the end is unattainable.

I have devoted an unusually long life to speculations which the present terminate ; and I claim, that
they constitute a better intellectual philosophy than
can be found elsewhere, and more intelligible. This
may not be saying much ; and succeeding investi-
gators will, I hope, accomplish more. For no person
is better assured than I, that an abundant scope exists
for a fruition of the hope. But I have attained my
own purpose ; for —

> While some in war and some in trade delight,
> My pleasure is to sit alone, and write.